Milk and Yogurt

Heinemann Library
Des Plaines, Illinois

Hazel King

© 1998 Reed Educational & Professional Publishing
Published by Heinemann Library,
an imprint of Reed Educational & Professional Publishing,
1350 East Touhy Avenue, Suite 240 West
Des Plaines, IL 60018

Designed by Celia Floyd
Illustrations by Sue Aldridge, Oxford Illustrators, p. 8 and Barry Atkinson, pp. 11, 13, 20, 21, 22, 23, 26, 28
Printed in Hong Kong / China

02 01 00 99 98
10 9 8 7 6 5 4 3 2 1

Library of Congress Cataloging-in-Publication Data

King, Hazel, 1962-
    Milk & yogurt / by Hazel King.
      p. cm. — (Food in focus)
    Includes bibliographical references and index.
    Summary: Describes how milk and yogurt are produced and how they are used for food in countries around the world. Includes several recipes and experiments.
    ISBN 1-57572-657-2 (lib. bdg.)
    1. Milk—Juvenile literature. 2. Yogurt—Juvenile literature.
3. Cookery (Milk)—Juvenile literature. 4. Cookery (Yogurt)—Juvenile literature. [1. Milk. 2. Yogurt.] I. Title.
II. Series.
SF250.5K55 1998
641.3'71—dc21
                         97-44110
                           CIP
                           AC

Acknowledgments
The Publishers would like to thank the following for permission to reproduce photographs:
Gareth Boden, pp. 4, 7, 16, 18, 19, 25, 27, 28; Trevor Clifford, p. 5; Robert Harding Picture Library, p. 21; Hulton-Deutsch, p. 6; National Dairy Council, p. 15; Tony Stone, pp. 10, 12 (Graeme Norways); Trip, p. 9 (H. Rogers); Zefa, p. 13.

Cover photograph: Trevor Clifford

Every effort has been made to contact copyright holders of any material reproduced in this book. Any omissions will be rectified in subsequent printings if notice is given to the Publisher.

Some words are shown in bold, **like this**. You can find out what they mean by looking in the Glossary.

Contents

Introduction

Milk is really quite amazing! It is the only form of nourishment a newborn baby needs, because human milk contains the correct balance of **nutrients** for a baby to survive the first weeks of life.

Most of us continue to have milk in our diet because we can use the milk of other mammals. Mammals are animals that feed their young with their own milk, just as humans do. Some of the mammals that produce milk are cows, sheep, goats, buffalo, reindeer, camels, llamas, and horses. Each animal produces milk that is exactly suited to the needs of its young.

A great deal of cows' milk is used today as a drink, in cooking, or to make other products such as butter, yogurt, cheese, and cream.

Some people find they are **allergic** to animals' milk so they might use **soy milk** instead, produced from the soybean. Because this milk comes from a plant, some vegetarians prefer it over animial milk.

In many parts of the world, milk from goats, sheep, and water buffalo is also used for drinking or making cheese.

Different types of milk and milk products

Milk is a versatile product and can be used in lots of different ways. As a drink, milk can be used for milk shakes, hot chocolate, or frothy cappuccino coffee. Both sweet and savory dishes can be made with milk which is why it is so useful. Sweet foods include ice cream, crème caramel, rice pudding and fudge. Even some cookies have milk added to them. Savory foods made with milk include lasagna, quiche, bread rolls, and some soups. Milk can also be used to make dishes that can be served as either dinner or dessert! Some examples are soufflés, omelettes, and pancakes.

For many years milk has been used to make other products. Cream, butter, cheese, and yogurt are all made from milk and are often called **dairy products**. These foods differ depending on the type of milk used to make them. For instance, you can get cows' milk yogurt and goats' milk cheese. Like milk, yogurt can be used in a variety of different ways and in both sweet and savory dishes.

A selection of products made from milk and yogurt

History of Milk

People have used milk from animals since the beginning of history. However, as milk turns **sour** easily, in the past it was impossible to keep it fresh. Milk turns sour because the **bacteria** it contains multiply very quickly and cause it to go bad, particularly if it is kept warm.

Some cattle diseases such as tuberculosis (TB) and brucellosis were passed to humans who drank milk that had turned sour. This caused the deaths of many people. Then, during the 1850's and 1860's, French scientist Louis Pasteur found that heating milk killed the bacteria. First he experimented with the bacteria in wine and then moved on to milk. He discovered that if milk was heated in the right way, all the disease-producing bacteria were destroyed.

The process developed by Louis Pasteur became known as **pasteurization** and it is still used to make our milk safe to drink. Today milk is pasteurized by heating it to 160°F (72°C) for 15 seconds and then, to keep it fresh, it is cooled quickly to below 50°F (10°C).

Louis Pasteur working in his laboratory

Although the milk we buy has been pasteurized, it must still be stored properly to keep it fresh. Milk and other **dairy products** should be stored in a refrigerator at a temperature of about 40°F (5°C). At this temperature bacteria multiply slowly, and milk can stay fresh for two or three days.

Milk does not have a strong flavor of its own. It should be kept covered in the refrigerator because it can absorb the odors and taste of other foods.

Bacteria

Bacteria are microscopic creatures so small that thousands of them can fit on a pinhead. Bacteria are all around you, all the time. Some are quite harmless but others are harmful if they are allowed to multiply inside you or if you eat the poisons they produce. In a warm room one bacterium can grow into several million within just 24 hours!

Some bacteria are useful. You will discover later that harmless bacteria are actually needed to make yogurt. Cheese is another dairy product made using bacteria.

Milk and cheese should be covered and stored in a refrigerator.

Around the World

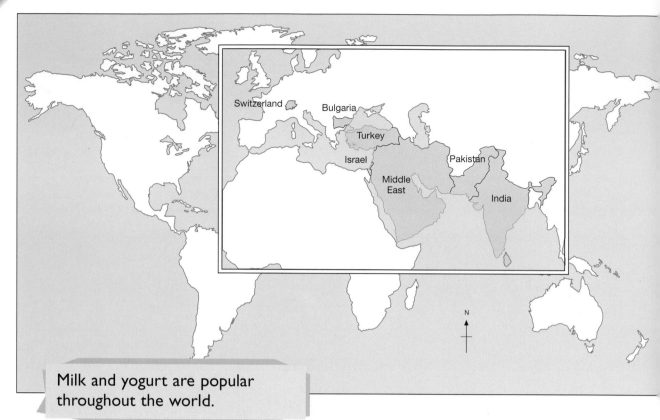

Milk and yogurt are popular throughout the world.

Both milk and yogurt are widely used throughout the world although the type may vary (cows', sheep's, goats' etc). Like all food products, they tend to be used in different ways in different places, depending upon the culture and traditions of the country. However, because people travel so much today, we are quite likely to use milk or yogurt in a way that is not traditional for our country.

Bulgaria

Yogurt is eaten a great deal here, which may be why Bulgarians are so healthy! It is used with salads, sauces, and desserts. Bulgaria is said to produce the world's best yogurt.

Switzerland

The breakfast cereal, muesli, was invented in Switzerland. It is served with milk or yogurt.

Turkey

A drink called *ayran*, made from yogurt and iced water, is popular here.

8

Israel

Many Jewish people live according to strict religious laws that include what kinds of food they are allowed to eat. One of these laws says that milk and meat must not be cooked or eaten together. Jews who strictly observe the laws must use two sets of cooking utensils and must wait three hours after eating a meal containing meat before eating anything containing milk or **dairy products**.

Pakistan

It is traditional to use yogurt in meat dishes in Pakistan and India. The meat is put in a **marinade** of yogurt and spices to make it tender.

India

Lassi is a refreshing yogurt drink that often accompanies Indian meals. *Raita* is made from yogurt, cucumber, and mint and helps to cool spicy curries. At weddings many special foods are served, including *shrikhand*, a spiced yogurt.

The Middle East

In this part of the world people regularly make yogurt at home, from cows' or sheep's milk. They believe that eating yogurt leads to a long and healthy life. A traditional Middle Eastern dish of stewed lamb is always served with yogurt.

Yogurt is used in many dishes from India.

Making Milk from Grass

Dairy cows are female cattle kept for the purpose of producing milk. They are only able to give milk once they have given birth to a calf.

A great deal of the cows' day is spent chewing grass, but they don't do it for fun—they need grass to produce milk. Cows actually have four chambers in their stomachs to break down and digest grass. First the cow uses its tongue to tear off the grass and mix it with saliva. Once the grass is swallowed, it is stored in the first two chambers—the reticulum and the rumen.

The grass hasn't yet been chewed so the cow will **regurgitate** the grass (known as cud) in small mouthfuls and chew and swallow it again. This is known as "chewing the cud."

In the rumen the grass is broken down by digestive juices. The water is then removed in the next chamber, the omasum, and what is left passes to chamber number four, the abomasum.

Cows spend much of their time grazing on grass.

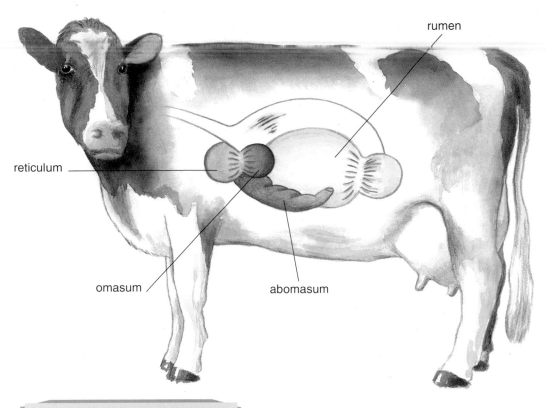

rumen

reticulum

omasum

abomasum

4 chambers of the stomach

More digestion takes place until the **nutrients** from the grass are absorbed into the cow's bloodstream where they can be used to keep the cow healthy and to produce milk.

The milk gradually builds up, drop by drop, in the cow's udder until it is full and the cow is ready to be milked. On average each cow provides 1,170 gallons (4,500 liters) of milk every year.

Allergic to cows' milk?

Humans are the only creatures that feed their young on another creature's milk. Babies are often fed on a special powdered milk made from cows' milk but altered to suit human babies.

However, some people discover they are **allergic** to cows' milk. It may cause a reaction such as eczema, asthma, diarrhea, arthritis, migraine, or catarrh.

Other people avoid cows' milk because they are "lactose intolerant." Lactose, a type of sugar found in milk, cannot be digested by everybody.

Instead of cows' milk, those with an allergy can have **soy milk**, goats' milk, or sheep's milk. Health food stores sell "dairy-free" products including margarine, bread, and cakes.

From Cow to Kitchen

Once a cow has chewed all that grass and made lots of milk, how does this become the milk you drink in your home or school?

1 Cows are kept by farmers who supply milk to dairies or to creameries for the manufacture of **dairy products**.
2 The cows have to be kept as clean as possible. They are milked twice a day, usually by machine, and the amount of milk each cow produces is recorded.
3 Milk from the cow passes to a refrigerated farm vat where it is cooled and stored. A tanker truck collects it the same day or the following morning.

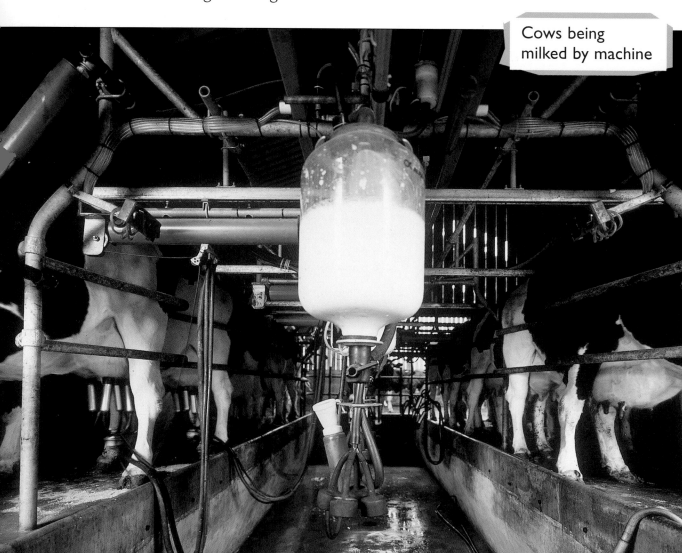

Cows being milked by machine

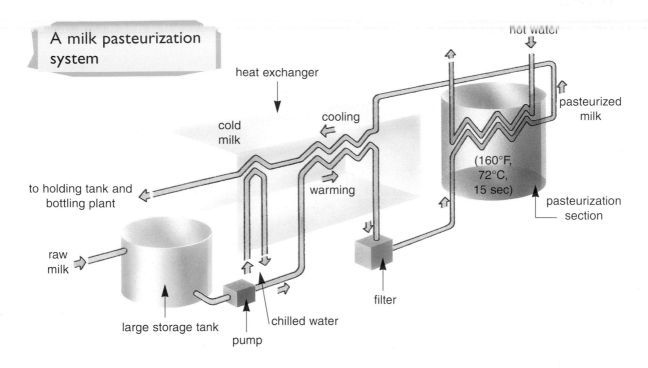

A milk pasteurization system

heat exchanger

hot water

cold milk

cooling

pasteurized milk

warming

(160°F, 72°C, 15 sec)

pasteurization section

to holding tank and bottling plant

raw milk

filter

large storage tank

pump

chilled water

4 Once milk arrives at a dairy, it is tested to make sure it is clean and free from disease. If it passes the test, the milk is unloaded into large storage tanks.

5 Next the milk is heat-treated using the **pasteurization** process, which destroys harmful **bacteria** but hardly affects the **nutritional value** or taste of the milk.

6 The milk may then be processed further before being hygienically bottled or packed into cartons or containers.

7 The milk is loaded into refrigerated trucks and taken to supermarkets, shops, and other retailers.

8 **Consumers** can buy milk in all sorts of stores from gas stations to supermarkets.

Cartons are filled with milk and the tops are sealed.

Making Yogurt

Yogurt can be described as "thickened milk" that has a sharp, tangy taste. Today it has a creamy, smooth texture very different from the yogurt produced in the past. Old-fashioned yogurt looked more like milk with lumps in it.

Just like milk, yogurt is versatile. It can be used for salads, sauces, desserts, breakfasts, in baking, as an **accompaniment** to a meal, and even as a drink.

As yogurt has a rather "healthy" image, food manufacturers have developed a huge range of different types and flavors. Unfortunately these are not always as "healthy" as they might seem because sugar and sugary foods are added to them, especially to those for children.

Yogurt manufacture

1 Yogurt is produced from whole milk or skimmed milk, depending on the type of yogurt to be made.
2 The yogurt is produced by adding a special **bacterial** culture, sometimes known as a "starter culture," to the milk. These bacteria are harmless. They are called *Lactobacillus bulgaricus*, after Bulgaria— where yogurt was first eaten.
3 The milk is heated to 110°F (43°C). At this temperature the bacteria cause the milk to set, or "clot," so that it looks a little like thick custard. It then has to **incubate** for several hours.
4 After this, the yogurt must be cooled very quickly to 41°F (5°C) to stop any bacteria from multiplying. It is stored at this temperature until it becomes slightly acidic.

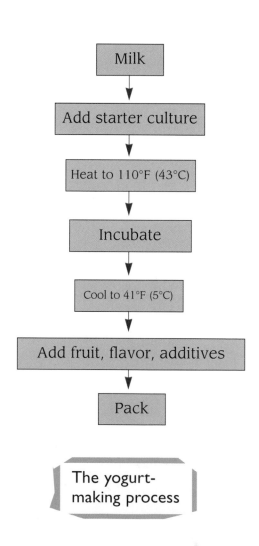

Milk

↓

Add starter culture

↓

Heat to 110°F (43°C)

↓

Incubate

↓

Cool to 41°F (5°C)

↓

Add fruit, flavor, additives

↓

Pack

The yogurt-making process

5 Additives may be used such as coloring, thickening, or **preservative.** Check the ingredients list! Other ingredients may include sugar, fruit puree or pieces, cereals, and nuts.

6 The yogurt is then poured into its container and the lid is sealed down. It must be stored at a temperature of 40°F (4.5°).

Sometimes the lid of a container of yogurt may bulge and the yogurt inside may taste fizzy or gassy. This happens with time as the fruit **ferments** and produces the gas carbon dioxide (CO_2). You should always eat yogurts before the use-by date on the carton.

Yogurt being produced in a factory

The finished product is packed up and ready to go to the stores.

What Type of Milk?

Today we can choose from a variety of milks. Each one has a useful purpose or is suited to a particular type of person. For example, fat free milk is available for people on low-fat diets while thick, sweet, condensed milk is excellent for making desserts, such as fudge.

Some of the different types of milk available

Whole milk

This is **pasteurized** milk that contains about 3.9% fat. Look carefully at a bottle of whole milk, and you will see that the fat or cream has floated to the top. This is because fat is lighter than milk.

Reduced fat milk

This is pasteurized milk with about half its fat removed so it contains only 1.5% to 1.8% fat. This means it is lower in **kilocalories (kilojoules)**.

Fat free milk

This is pasteurized milk as a low-calorie food, with virtually all its fat removed, leaving just 0.1% fat. It is ideal for someone wanting to lose weight; however, it is unsuitable for babies and young children.

Homogenized milk

To produce this type of milk, pasteurized whole milk undergoes a special process. The milk is forced through tiny holes to break up the fat. When it is left to stand, the fat does not rise to the top but remains floating throughout the milk.

Sterilized milk

In this process, milk is heated to 235°F (113°C) and then homogenized. This gives it a slightly different flavor from nonsterilized milk. If unopened it will keep for several months.

Ultra heat-treated milk (UHT)

UHT milk is a type of **sterilized** milk. It is heated to 270°F (132.2°C) for one second, then it is poured into special cartons. Like sterilized milk, if unopened it will keep for several months.

Evaporated milk

This is sterilized milk that has been concentrated or made thicker by **evaporation** (which removes some water). It is sold in cans and will keep unopened for many months.

Condensed milk

This is also concentrated milk but with sugar added to help **preserve** it. Whole, reduced fat, or fat free milk is used. It is very thick and sweet.

Dried milk powder

Dried milk powder is a useful long-storage "stand-by." It is made from whole or fat free milk, which is homogenized, heat-treated, and dried to remove all the water. When you need milk, you simply mix the dried powder with water to produce liquid milk.

Flavored milk

Different types of flavored milks are made from long-life (UHT or sterilized) or fresh milk. Flavors include chocolate, strawberry, or banana.

What Type of Yogurt?

Yogurt is popular with both adults and children. This is probably because food manufacturers have come up with so many different varieties. You can buy yogurt in all sorts of flavors; the containers come in all shapes and sizes, and you can even add your own flavors.

There are thick and creamy yogurts, low-fat yogurts, set yogurts, even some made especially for breakfast. Here are just a few of the yogurts available.

Low fat, very low fat, diet, or light yogurts

These are made using fat free milk and may be natural or flavored. Flavored yogurts may have sugar added, they many not be as healthy as they sound.

Bio yogurts

By adding extra cultures, a mild-tasting yogurt is produced. These are thought to be particularly healthy as they are supposed to assist the digestion.

Set or French yogurts

These yogurts are heat-treated and sealed in the container, where they are left to **ferment** and set. They have a more solid texture than other yogurts.

A few varieties of yogurt

Thick and creamy yogurts

Whole milk is used to make these yogurts, and some of them have cream added as well. They may be natural or flavored.

Greek or Greek-style yogurts

For a rich flavor choose a Greek yogurt! These are made with whole milk and have a high fat content.

Children's and infants' yogurts

To attract children these yogurts are usually sold in cartons covered with cartoon characters. They usually contain a smooth puree of fruit rather than pieces. Infants' yogurt should have a low sugar content. Children's yogurt is available with (unhealthy!) sugary foods that can be stirred in.

Do-it-yourself yogurt! Add the flavor to suit your taste

Split or corner yogurt

Just to make eating more fun, you can now buy a container of yogurt with a separate corner containing a flavoring. You can add this to the yogurt if and when you please! The flavoring might be fruit puree, cereal grains, or chocolate-coated cereal.

Fromage frais

Although these can usually be found with the yogurts and other cold desserts, in fact they are a form of soft cheese. Eaten as a dessert, they come in small cartons containing a very creamy product that may be plain or lightly flavored. They were first eaten in France.

But Is it Good for You?

When people say milk is "good for you," they are referring to is its **nutritional value**. The nutritional value of a food depends on the type and amount of **nutrients** it contains. Nutrients are vitamins, minerals, and other building blocks used by our bodies to stay alive and healthy.

Nutrients

The nutrients found in milk

protein

energy

vitamin B₁ (thiamin)

vitamin B₂ (riboflavin)

phosphorus

vitamin A

vitamin D

vitamin B₁₂

fat

calcium

Yogurt contains the same nutrients because it is made from milk but the amounts will vary. Different types of milk and yogurt will also contain different amounts of each nutrient.

Why are these nutrients important?

Energy	All food provides energy, which is measured in **kilocalories** or **kilojoules**. Our body burns this energy all the time to keep us functioning properly.
Protein	This is needed for the formation, growth and repair of our body tissues.
Fat	Fat supplies us with lots of energy and helps our body to absorb vitamins. It also protects our **internal organs**.
Calcium and phosphorus	These are both minerals that work together to keep our bones and teeth strong.
Vitamin A	This vitamin is important for our eyes as well as for growth and a healthy skin.
Vitamin D	This vitamin is only found in small quantities in milk although you can buy varieties with "vitamin D added." It is required for our growth and development and for the formation of bones and teeth.
Vitamin B₁ Vitamin B₂	Both are needed to help the body release the energy that food supplies.
Vitamin B₁₂	This is essential for our growth and the formation of red blood cells and for our nervous system.

This girl is enjoying a glass of milk.

Milk and yogurt are found in the "Eat moderately" section of the food pyramid. This is because they contain protein and some fat as well as vitamins and minerals. We should try to eat a moderate amount every day. For example, a twelve-year-old boy might have milk on his cereal, a milk shake at lunchtime, and a glass of milk with his evening meal.

The food pyramid

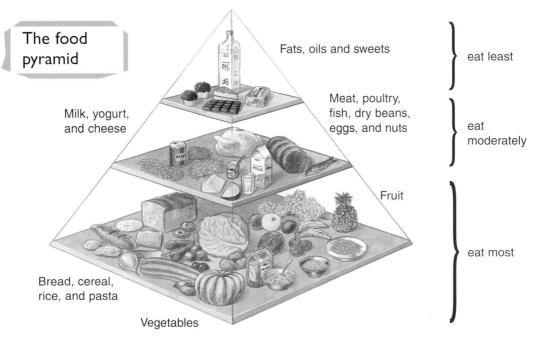

Fats, oils and sweets — eat least

Milk, yogurt, and cheese

Meat, poultry, fish, dry beans, eggs, and nuts — eat moderately

Fruit

Bread, cereal, rice, and pasta

Vegetables — eat most

Experiment with Milk

Using milk to make sauces or drinks often involves heating it in a saucepan. When milk is heated, several changes occur. This experiment will help you to understand these changes. You must be very careful when using the stove or hot plate. Ask an adult to help you before you start.

Heating milk

Read through the experiment before you begin so that you know what to do.

Watch what happens very carefully!

You will need:

- about 1¼ cup (300 ml) milk
- 1 heavy-bottomed saucepan
- a stove or hot plate
- 1 small bowl

What to do:

1 Pour the milk into the saucepan.
2 Heat the milk gently and watch very carefully. You should be looking out for:

- steam rising from the top of the milk
- a skin forming that may look wrinkled
- tiny bubbles around the edges of the saucepan and underneath the milk's skin
- the milk rising up the sides of the saucepan.

3 Take the saucepan off the heat *before the milk boils over*.
4 Let the milk cool for a few minutes then pour it into the bowl.

What happened?

There are good reasons why all those changes took place. They could only happen because milk contains protein, fat, calcium and water.

When milk gets hot some of the water in it turns to steam. This is what you could see rising from the top of the milk. Because fat is lighter than milk, any fat will float to the surface. Did you notice it become a creamy color?

Then the protein changes. Like all proteins, it "sets" when it is heated, whichis what creates the skin on the milk.

Now the milk has a "lid" on it. As it boils, the bubbles of steam underneath push the skin and force it upwards, making the milk look like it is growing.

This is what makes milk boil over so easily.

After pouring out the milk did you notice a layer on the bottom of the saucepan? Heating causes some of the protein and calcium in the milk to fall to the bottom where it settles. This layer will burn if you do not use a heavy-bottomed saucepan.

Watch out! Milk boils over easily.

23

Recipe: Milk Shakes

Ice-cold milk shakes are nutritious and delicious, especially when you make them yourself. They are popular with children and are drunk in most countries in the western world.

You do not have to use a blender to make these shakes but if you do, ask an adult to help you. You can serve milk shakes in tall glasses add decoration to make them look special. Each recipe serves two people.

Banana shake

Serves 2

You will need:
Ingredients
- 8 oz. (227 g) container natural yogurt
- 2 medium bananas
- 1 1/4 cups (300 ml) fresh milk
- sprinkling of nutmeg for decoration

Equipment
- blender (or hand whisk and a large bowl)
- 2 tall glasses

What to do:

1 Place the yogurt, bananas, and milk in the blender and close the lid firmly. Blend for 30 seconds. (Or hand whisk the ingredients together in bowl.)
2 Pour into two glasses and sprinkle lightly with nutmeg.

Chocolate shake

Serves 2

You will need:

Ingredients
- 1 heaped teaspoon chocolate powder or syrup
- 1 1/2 tablespoons boiling water
- 2 cups (450 ml) fresh milk
- 2 heaping tablespoons (30 ml) vanilla ice cream

Equipment
- bowl
- teaspoon
- tablespoon
- balloon whisk
- 2 tall glasses

What to do:

1 Place the chocolate powder or syrup and water in the bowl. and mix to a smooth paste with the teaspoon.
2 Pour in the milk and whisk together using the balloon whisk.
3 Add 1 tablespoon (15 ml) of vanilla ice cream and whisk again.
4 Place the remaining ice cream in the bottom of the glasses. Pour the milkshake over the ice cream.
5 Serve immediately.

Sunshine shake

Serves 2

You will need:
Ingredients

- 1 1/4 cups (300 ml) fresh milk
- 2/3 cup (150 ml) pineapple juice, chilled
- 2/3 cup (150 ml) orange juice
- 1 teaspoon honey
- 2 slices of orange and sprigs of mint for decoration

Equipment

- blender (or hand whisk and a large bowl)
- 2 tall glasses

What to do:

1 Whisk or blend all the ingredients together.
2 Pour into the glasses and decorate with slices of orange and sprigs of mint.

Strawberry shake

Serves 2

You will need:
Ingredients

- 8 oz (227 g) strawberry yogurt
- 1 1/4 cup (300 ml) fresh milk
- strawberry for decoration

Equipment

- blender (or hand whisk and a large bowl)
- 2 tall glasses

What to do:

1 Whisk or blend all the ingredients together.
2 Pour into the glasses and decorate with sliced strawberries.

Ice cold, home-made milk shakes

Recipe: Hot Baked Sandwiches

Sandwiches are a well-known snack invented by the British Earl of Sandwich. The original sandwich has been changed and developed many times over the years. Why not try this version—sandwiches hot from the oven! Ask an adult to help you.

Just like ordinary sandwiches, baked sandwiches can be made with many different fillings. For instance, if you do not like cheese and tomato you could use chopped bacon and sliced mushrooms. If you do not like relish, use tomato ketchup, steak sauce, mustard—whatever you like!

Cheesy sandwiches

Serves 1

You will need:

Ingredients

- $7/8$ oz (25 g) cheese, e.g. cheddar or American
- $1/2$ tomato
- 2 slices bread
- teaspoon of relish (optional)
- a little butter or margarine
- 2/3 cup (150 ml) milk
- 1 egg
- black pepper (optional)

Equipment

- cutting board
- sharp knife
- spreading knife
- shallow ovenproof dish, approximately 6 x 4 inches (15 cm x 10 cm)
- measuring cup
- fork, spatula
- oven mitts
- serving plate

What to do:

1 Make sure the rack is in the middle of the oven. Preheat the oven to 350°F (180°C).
2 Place the cheese on the cutting board and slice into thin pieces. Slice the tomato.
3 Place the bread on the board and lay the cheese on one slice. Top with tomato slices. Spread the other slice of bread with pickle relish or mustard.
4 Place the bread with relish on top of the other piece, to make a sandwich. Lightly spread some margarine or butter over the top of the sandwich then carefully cut it into four triangles.
5 Rub a little butter or margarine around the bottom and sides of the dish then place the sandwich triangles in it. Measure the milk into the measuring cups. Break the egg into the milk and add a sprinkling of pepper, if desired. Whisk thoroughly using a fork. Pour this mixture over the sandwich in the dish.
6 Using oven mitts, carefully place the dish in the oven on the middle shelf. Bake for 25 minutes. At the end of the cooking time, the sandwich should be lightly browned and the egg mixture will have set.
7 Carefully remove the dish from the oven, using oven mitts. Place on a heatproof surface and allow to cool for one minute. Run a knife around the edges then carefully lift the sandwiches out with a spatula. Put on a plate and serve.

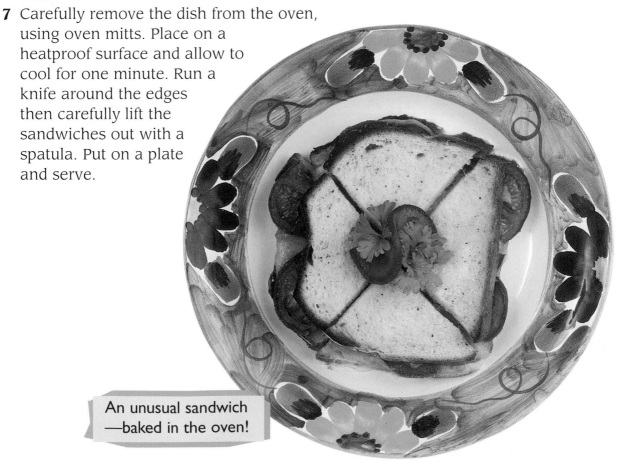

An unusual sandwich —baked in the oven!

Recipe: Tilted Parfaits

These desserts look really interesting because the jello layer is tilted at an angle.

Traditionally jello is popular in America, Australia, and Britain and is especially popular with children! Dessert jelloes are usually fruit flavored and when they are ready for eating they have a special wobbly feel.

These treats taste as good as they look!

Tilted parfaits

Serves 4

You will need:

Ingredients

- water
- 1 package of your favorite flavor jello
- fresh fruit such as a large banana, pear, or peach, or about 5 1/4 oz (150 g) raspberries, strawberries, blueberries
- 1 package instant vanilla pudding
- 1 tablespoon sugar
- 2 1/2 cups (600 ml) lowfat milk
- 4 tablespoons yogurt
- chocolate chips for decoration

Equipment

- tea kettle
- measuring cup
- knife
- cutting board
- 4- 8 oz (300 ml) glasses or tall sundae dishes
- tablespoon
- small bowl
- saucepan
- teaspoon
- wooden spoon

What to do:

1 Fill the tea kettle and leave it to boil. Place the jello in the measuring cup. Carefully follow the directions on the package and pour in boiling water.

2 Stir until the jello is dissolved. Following the directions, add cold water. Stir again.

3 If necessary, wash the fruit and remove the stalks or peel. Chop into small pieces, then divide among the four glasses.

4 Pour 2/3 cup (150 ml) of jello over the fruit in each glass. Carefully prop the glasses in the refrigerator at an angle so the jello can set tilted in the glass. Leave to set.

5 Make the vanilla pudding by following the directions on the package.

6 When the pudding is cool and the jello is set, stir the yogurt evenly into the pudding. Take the glasses out of the refrigerator and pour pudding on top of each jello. Chill in the refrigerator.

7 Decorate each tilted parfait with chocolate chips when you are ready to serve them.

Glossary

accompaniment a food served with other foods because they go well together e.g. cool yogurt to accompany a spicy curry

allergic a person is said to be allergic to something if it causes them to have an unpleasant reaction—for example, some people are allergic to dairy products: if they eat them, they might get a rash on their skin or suffer from a bad headache

bacteria microscopic creatures that multiply quickly in warm conditions but are destroyed at high temperatures. Some are harmless but some can cause diseases or death

consumers people who buy products

dairy products food products made from milk—cheese, yogurt, cream, and butter

evaporation removing water from substances by heating

ferment a chemical reaction in which carbon dioxide gas is produced, e.g., a fruit yogurt that has been kept too long may go fizzy or gassy

incubate to encourage the development of bacteria by keeping a substance warm, e.g., during yogurt making

internal organs hollow or solid organs inside the body such as liver, kidney, and heart

kilocalories traditional units used in measuring the energy in our food e.g., one pint of whole milk provides about 375 kilocalories

kilojoules modern units used in measuring the energy in our food, e.g., one pint of whole milk provides about 1,632 kilojoules (1 kilojoule = 4.2 kilocalories)

marinade a flavored liquid in which food such as meat is soaked before being cooked, e.g., chicken may be left in a marinade made from natural yogurt and spices before being cooked as a main meal

nutrients protein, carbohydrate, fat, minerals and vitamins contained in food

nutritional value the type and quantity of nutrients in a food

pasteurization the heat treatment of milk to destroy bacteria

preservative something added to food to preserve it—this may be another food, e.g. sugar, or it may be a chemical preservative, e.g., ascorbic acid, which is sometimes added to fruit yogurts

preserve   to treat a food so that it keeps longer than it would naturally, e.g., UHT milk is preserved by being heated so that it lasts longer than fresh milk, if unopened.

regurgitate   bringing swallowed food back up into the mouth (something cows do naturally)

sour   to go bad. When milk turns sour it is no longer fresh, and it has an unpleasant taste and smell.

soy milk   "milk" made from soy beans—popular with people who are allergic to animals' milk

sterilized   treated to be made free from bacteria

More Books to Read

Cooper, Jason. *Dairy Products.* Vero Beach, FL: Rourke Publications, 1997.

Fitzsimons, Cecilia. *Dairy Foods & Drinks.* Morristown, NJ: Silver Burdett Press, 1997.

Powell, Jillian. *Milk.* Chatham, NJ: Raintree Steck-Vaughn, 1997.

Index